ANIMAL WORLD

THE SMARTEST ANIMALS IN THE WORLD

by Clara MacCarald

BrightPoint Press

San Diego, CA

© 2024 BrightPoint Press
an imprint of ReferencePoint Press, Inc.
Printed in the United States

For more information, contact:
BrightPoint Press
PO Box 27779
San Diego, CA 92198
www.BrightPointPress.com

ALL RIGHTS RESERVED.

No part of this work covered by the copyright hereon may be reproduced or used in any form or by any means—graphic, electronic, or mechanical, including photocopying, recording, taping, web distribution, or information storage retrieval systems—without the written permission of the publisher.

LIBRARY OF CONGRESS CATALOGING-IN-PUBLICATION DATA

Names: MacCarald, Clara, 1979--author.
Title: The smartest animals in the world / by Clara MacCarald.
Description: San Diego, CA: BrightPoint Press, [2024] | Series: Animal world | Includes bibliographical references and index. | Audience: Ages 13 | Audience: Grades 7-9
Identifiers: LCCN 2023008709 (print) | LCCN 2023008710 (eBook) | ISBN 9781678206185 (hardcover) | ISBN 9781678206192 (eBook)
Subjects: LCSH: Animal intelligence--Juvenile literature. | Animal behavior--Juvenile literature.
Classification: LCC QL785 .M165 2024 (print) | LCC QL785 (eBook) | DDC 591.5/13--dc23/eng/20230327
LC record available at https://lccn.loc.gov/2023008709
LC eBook record available at https://lccn.loc.gov/2023008710

CONTENTS

AT A GLANCE

- Smart animals use their intelligence to help them survive.

- Chimpanzees make and use many different kinds of tools. For example, they prepare sticks to fish for tasty termites and use rocks to break open nuts.

- Chimps communicate with each other using sounds, facial expressions, and gestures.

- African gray parrots can mimic human language. One parrot learned to name fifty objects, seven colors, and five shapes.

- African grays can solve puzzles and work together to get treats.

- Each bottlenose dolphin has its own distinct whistle. Dolphins can identify each other by their whistles.

- Some bottlenose dolphins hunt with tools. They protect their beaks with sea sponges when they dig, and they catch fish using empty seashells.

- Coconut octopuses carry coconut shells around. The octopuses cover themselves with the shells to hide from predators and sneak up on prey.

- Octopuses in captivity can cause a lot of mischief. They sometimes squirt water at people and mess with their tanks to cause flooding.

A TOOL FOR TERMITES

A chimpanzee walks through a forest in the Republic of the Congo. It holds a stick in one hand. The stick is a tool that the chimp made. First, the chimp pulled a plant stem out of the ground. Then the chimp tore a leaf off the end of the stem. Now the chimp plans to use the tool to get food.

After a short walk, the chimp arrives at a termite mound. The mound looks like a small hill. Chimps often come here to eat the tasty termites that live inside.

The chimp needs to give its tool a finishing touch. The animal chomps down on the end several times. The stem

breaks apart. Now the thin bits of wood look like a brush. The tool is ready to use.

Along the mound are filled-in holes made by the termites. The chimp chooses one. It opens the hole with its hand and sticks the tool inside.

The termites within the mound bite onto the stem. The chimp yanks the stick out. Several termites cling to the end. The chimp eats them, enjoying a satisfying snack.

ANIMAL INTELLIGENCE

Smart animals use their intelligence to help them survive. Animals can use tools.

They can solve problems. Many animals communicate with each other. Some work together in intelligent ways. While there are a lot of smart animals, a few animals in particular are known for their brainy skills.

1
CHIMPANZEES

Chimpanzees are among the great minds of the animal world. The chimp is a **species** of ape. It's also one of the animals most closely related to humans. In the wild, chimps make their homes in the forests and grasslands of Africa. Chimps live together in large groups. They eat mostly plant matter such as fruits. But they can eat other things too. Their diet can include insects, eggs, and meat.

Chimps show their smarts in several different ways. They use tools,

communicate with each other, and solve

problems. They also learn from each other.

Because of this, one group of chimps may

have a different **culture** than another.

For example, each group of chimps

might have a different way of making or
using tools.

CHIMPS AND THEIR TOOLS

Lots of animals use tools, but chimps excel
at it. Sometimes chimps throw sticks or
stones to scare other animals. Chimps
can use leaves to wipe their mouths clean.
They also use tools to get food and water.
For example, they use stones to crack
open nuts.

Sometimes chimps make their own
tools, such as how chimps chew the ends
of sticks to catch termites. Chimps use a

similar process to make spears. The apes break branches off of trees and use their teeth to sharpen the ends. They use the spears to hunt small animals.

Chimps can also combine different tools. Sometimes a chimp uses a wad of leaves like a sponge. It puts it in a tree to collect water. But sometimes the chimp puts the leaves too deep. If it can't reach the leaves, the chimp uses a stick to push them out.

Some chimps make sets of tools for gathering honey. These sets include several separate parts. The chimps must use each tool at the right time. Chimps

first poke thin sticks into the ground to
search for underground bee nests. When
they find one, they use a big stick called a
pounder to break it open. They then use
smaller sticks to make the opening bigger.
Chewed sticks and bark strips help the
chimp pull out the honey.

STONE AGE CHIMPS

Chimpanzees have been using tools for a long
time. Scientists have found stone hammers made by
chimps 4,300 years ago. The tools were too heavy
for humans but the right size for chimps. Like today's
chimps, ancient chimps used hammers to break
open nuts.

Like humans, chimpanzees are all different. Each chimp has a unique personality.

Chimps make some of these tools at the hive. They make other tools ahead of time. That takes planning. It takes intelligence for an animal to know what it will need in the future.

SOCIAL KNOW-HOW

Communication is another way that animals show off their mental skills. Communication is when one animal signals to another. Chimps can make a wide range of sounds as signals. For example, they bark to invite other chimps to join in a hunt.

Chimps can also communicate across long distances using the huge roots of rain forest trees. Chimp scientist Catherine Hobaiter explains, "If you hit the roots really hard, it . . . makes this big deep, booming sound that travels through the forest."[1] Individual chimps have their own patterns of

Chimps help keep each other clean. A chimp may help its friend by picking dirt and bugs from the other chimp's hair.

drumbeats and sounds that they can use if

they want their identity to be known.

Chimps also communicate with **facial**

expressions. Some of these movements,

such as grinning, look familiar to humans.
Chimps also use **gestures**. Chimps may
offer a friend their shoulder to let the friend
know they want to be groomed.

Chimps can even use their smarts to trick
each other. Some chimps hide their facial
expressions so a stronger chimp doesn't
see their fear. Chimps can also sneak
around to get food so they don't have to
share with others.

TEST TAKERS

Scientists have performed many tests
to see how intelligent chimps are.

Some experiments require the animals to use a tool. The chimps might have to choose between a working tool and a broken one.

One classic test is called the mirror test. Scientists give medicine to the animal to put it to sleep. Then scientists paint a spot on the animal's body somewhere it can't

JANE GOODALL

In 1960, Jane Goodall began studying chimps in the wild. At the time, scientists didn't think animals used tools. Goodall was the first to discover a chimp using a stick to fish for termites. Since then, Goodall has spent much of her life teaching people to understand and protect chimps.

normally see. When the animal wakes, the scientists show the animal a mirror.

To pass the test, the animal needs to notice the spot on its body. This shows that the animal recognizes the image in the mirror as itself. Many species don't pass the mirror test. Chimps are among those that can. Failing doesn't prove an animal lacks intelligence. But passing shows that chimps are aware of themselves in ways that some animals are not.

2

AFRICAN GRAY PARROTS

The African gray parrot is a large bird that may rival the chimp in intelligence. In the wild, the parrots live in and near forests in Central and West Africa. At night, they gather in flocks with as many as 1,000 members. During the day, smaller groups forage for fruit, seeds, and nuts.

There are more than 350 species
of parrots. Parrot species include
macaws, cockatoos, and parakeets.

There are two types of African gray parrots: Congo grays and Timneh grays. Congo grays (pictured) are bigger than Timneh grays. They also have lighter feathers and darker beaks.

Parrots are smart. They can use tools.
They can solve problems. People have kept
parrots as pets since ancient times because
the birds can learn and perform tricks. Their
most famous tricks involve sound.

In the wild, these social birds
communicate with a range of calls. Many
parrot species are excellent at **mimicking**
the noises made by other species. Several
are known to mimic human speech. But
of all parrots, African grays do the best
job sounding like humans. This ability has
helped scientists test just how smart these
birds are.

PARROTS AND LANGUAGE

Research has shown that African grays can

learn to understand the words they mimic.

Biology professor David Aborn explains

why mimicking is important for parrot

research. "Parrots can be trained to speak

our language," Aborn says. "They can respond in ways we can understand."[2]

Alex was one of these trained parrots. Alex learned to use about 150 English words. He could label objects based on traits such as color and shape. He could tell similar objects apart by color and material.

Alex could name fifty objects, seven colors, and five shapes. He could count to six. Alex could even put words together to talk about something he hadn't seen before. He could ask for objects that were out of view.

THINKING LIKE A BIRD

Scientists have shown that African grays can use reasoning to solve puzzles. For example, one set of tests used objects hidden under cups. The simplest test used two cups. A scientist hid a treat. Then the person showed a parrot that one cup

was empty. The parrot correctly guessed

the treat was under the other cup. Other

tests were harder and used several cups.

The parrot was still able to figure out where

the treat was.

African grays have very good memories

too. In a different test, a parrot watched

a scientist hide different-colored objects under cups. The scientist moved the cups around. Then the scientist asked the parrot to find a specific object. The parrot did a better job remembering which cup held the correct object than children did. The parrot even did better than adult humans on some tests.

African grays also display self-control. There is a classic test of self-control called the marshmallow test. Scientists created this test for young human children. They give a child a marshmallow. The child can eat the marshmallow right away. But a child who waits to eat the treat will be given a second marshmallow.

The test for African grays was a little different. Scientists gave the birds sunflower seeds on a rotating tray. If the parrots didn't eat the seeds, the tray would keep rotating and the birds could have walnuts.

African grays like walnuts a lot better than sunflower seeds. On average, the parrots could wait about thirty seconds before eating the sunflower seeds. One parrot waited for fifty seconds. To keep themselves busy, some birds walked around or played with a toy.

HOW MANY WORDS CAN ANIMALS LEARN?

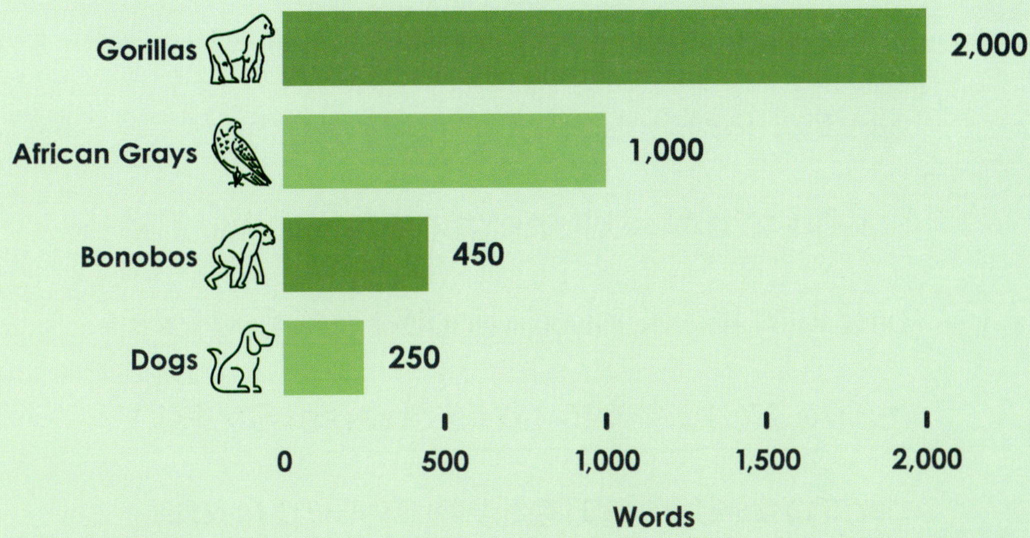

Sources: Brooke Borel, "Talking with Animals: Seven Examples of Interspecies Communication," TED Blog, June 6, 2013. https://blog.ted.com.

"Smarter Than You Think: Renowned Canine Researcher Puts Dogs' Intelligence on Par with 2-Year-Old Human," American Psychological Association, 2009. www.apa.org.

Many animals are smart enough to understand words. Some can even respond.

These parrots help each other in smart ways. In one test, two African grays were in cages next to each other. Scientists gave the birds objects. The parrots learned they could trade the objects through a window for treats. But sometimes the scientists blocked one of the windows. The bird with the blocked window seemed to realize it couldn't trade its objects. So the bird passed the objects to the bird next door. The other bird traded the objects for food instead.

African grays can also work together. In some tests, the birds had to pull a string at the same time to get treats. In another, one bird had to sit on a seesaw while the other pulled a string to release the treats.

Scientists still have more to learn about these smart birds. "It's not that we proved everything provable," scientist Irene Pepperberg says about her research with the species. "It's that we've [shown] a behavior that leads to a lot of different questions."[3] Scientists continue to try to answer these new questions.

African gray parrots are listed as endangered by the International Union for the Conservation of Nature. One threat the birds are facing is wildlife trafficking. Since the 1980s, more than 1.3 million African grays have been captured for the pet trade.

3
BOTTLENOSE DOLPHINS

Dolphins are well-known for their intelligence. There are thirty-six dolphin species in the world. Bottlenose dolphins are perhaps the most famous. Many scientists have studied bottlenose dolphins both in **captivity** and in the wild.

Bottlenose dolphins live all around the
world in oceans and rivers connected to the
sea. Dolphins are very social. They often
form groups called pods. Individuals can
move around between pods.

*Bottlenose dolphins live for about forty-five to fifty
years in the wild.*

Bottlenose dolphins hunt animals such as fish, shrimp, and squid. These predators show a lot of their smarts in how they go after their prey. Dolphins learn from each other. They can work together. Some wild dolphins even learn to work with humans to get food.

DOLPHIN TALK

Dolphins communicate with each other by using sounds. They make whistles and squeaks. In the wild, bottlenose dolphins have something similar to a name. Each individual has its own specific whistle.

When a dolphin whistles, the other dolphins know who is nearby.

Dolphins can also mimic each other's whistles. The sound might catch the other dolphin's attention, like calling someone's name. Dolphins can even remember their friends. Scientists found that some dolphins could recognize an individual's whistle after not seeing the other dolphin for more than twenty years.

DOLPHIN HUNTS

Dolphins are very skilled at hunting with echolocation. Echolocation is the use of sound to find things. A dolphin makes a noise that sounds like clicking. The noise bounces off an object or animal. Hearing the echo lets the dolphin figure out where an object is, what it is, and if it's moving.

Baby bottlenose dolphins are called calves. Calves stay with their mothers for three to six years.

Some dolphins use tools for hunting. In Shark Bay, Australia, dolphins use basket sponges. Basket sponges grow on the seafloor. A dolphin breaks one off. It puts its beak inside the sponge. The dolphin then searches for fish buried in the rocks and sand. The sponge protects the dolphin's nose as the animal pokes around. Mother dolphins teach this trick to their young.

Dolphins also use empty shells to hunt fish. The shells come from large sea snails. Scientist Sonja Wild explains how the trick works. She says that a dolphin will chase a fish into one of the shells. The dolphin then sticks its beak inside the shell to pick it up. The animal then holds it over the water and shakes it. Wild says, "The fish basically falls into their open mouth."[4]

Sometimes dolphins hunt together. A pod of dolphins will find a school of fish in shallow water. One dolphin swims in rapid

Members of a dolphin pod hunt together, play with each other, and protect each other.

circles around the school. The dolphin's tail

stirs up mud. The rising wall of mud traps

the fish and scares them. The fish jump out

of the water and over the mud to escape.

Other dolphins wait outside the circle to

catch them.

Dolphins have been known to rescue people in danger. Some have saved people from drowning. Others have protected people from sharks.

DOLPHINS AND HUMANS

Dolphins have also learned how to hunt with humans. In Brazil, there are groups of dolphins that work with local fishers. The humans wait near shore. The dolphins herd fish toward them. When the dolphins get close, they slap the water. This is a signal for the fishers to cast their nets. The fish scatter, making it easier for the dolphins to grab a meal.

Dolphins are very playful with each other and with humans. Scientist Denise Herzing explains a game dolphins play if they get a toy. "They like to let you get almost close enough to grab the toy," she says, "but then they speed off."[5]

Dolphins can learn to understand hand gestures. Some scientists made up a language of hand signs. The signs gave the dolphins commands to do tricks. The dolphins could watch short sentences

DOLPHINS IN THE NAVY

Scientists aren't the only ones interested in dolphin smarts. Soldiers are too. The US Navy has used dolphins since 1959. Dolphins in the Navy can watch out for enemy swimmers. The animals perform better than machines at finding objects such as mines underwater. They can also recover lost gear.

made up of the signs and then follow them

in order.

In captivity, humans can teach dolphins

to perform many tricks. Dolphins can even

learn tricks from each other. For example,

one dolphin named Billie spent a short

period of time in captivity. She learned

a trick called tail walking. Tail walking is

when a dolphin beats its tail to hold most

of its body upright out of the water. People

returned Billie to the wild. Billie continued to

do the trick, and other dolphins around her

started tail walking after watching her.

4
OCTOPUSES

There are about 300 octopus species. Some are a few inches long, while others can reach 18 feet (5.5 m). Octopuses are mollusks. Mollusks are a group of animals with soft bodies and no bones. The group also includes snails, slugs, and clams. Clams aren't known for their smarts.

But the octopus is one of the great minds of

the animal world.

Octopus brains are very different from

those of chimps and other smart animals.

Most octopus **nerve cells** aren't even

in the creature's brain. They're in the

octopus's arms. Yet octopuses show
several signs of intelligence. They are playful
and curious. They can also cause mischief.
Scientists have even discovered that
octopuses appear to dream.

MASTERS OF HIDING

Octopuses have many tricks to prevent
their predators and prey from spotting
them. They can **camouflage** themselves.
Octopuses do this by changing how their
skin looks. They can change colors. They
can make their bodies appear bumpy like a
sea plant.

Octopuses can also mimic other animals.

The mimic octopus is a species that is

particularly good at mimicking fish. It can

move along the seabed like a flatfish.

Predators avoid flatfish because the fish can

be poisonous. The mimic octopus can also look like a sea snake or a lionfish.

Veined octopuses are famous for using tools. Veined octopuses use coconut shells as shelters for hiding. But the animals don't just dart into the shells to escape. They do something much smarter. They gather the shells for future use. The animals appear to be thinking ahead.

An octopus starts by finding half shells and cleaning mud out of them. The animal then stacks the shells. What octopuses do next surprised the scientists who first observed the behavior. "I could tell that the

Octopuses can hide in coconuts, seashells, and human trash.

octopus . . . was up to something," says Julia Finn, one of the scientists. "But I never expected it would pick up the stacked shells and run away."[6]

The animal moves along the seafloor while holding the shells. When a predator appears, the octopus builds a shelter.

It puts two shells together and hides in the

middle. Using its arms, the octopus pulls

the shells tightly together. All the predator

sees is a coconut. An octopus can also

hide from a prey animal before popping out to grab a meal.

Other octopuses use stones to hide. Octopuses can create dens out of stones. Some may even use stones as doors.

A CURIOUS MOLLUSK

Octopuses are curious about the objects and creatures around them. They're good at figuring things out. Around humans, octopuses can cause all sorts of mischief. For example, anyone trying to record an octopus should beware. Octopuses are known to open up underwater cameras.

If water gets inside the camera, the camera is ruined.

The animals can cause trouble in captivity too. With their clever arms and lack of bones, octopuses are good at escaping. Some octopuses have learned to squirt water at light bulbs to break the bulbs. Octopuses appear to recognize people.

THE CLEVER CUTTLEFISH

Cuttlefish are intelligent mollusks related to octopuses. Cuttlefish can solve puzzles and mazes. One study even gave them a version of the marshmallow test. Cuttlefish were able to stop themselves from eating a piece of king prawn. They used self-control to wait and get a tastier grass shrimp later.

If an octopus decides it dislikes a person, the animal might spray the person with water. Octopuses may also mess with their tanks to cause flooding.

To keep octopuses in captivity busy, scientists give them puzzles and toys. These puzzles and toys look different from the ones humans use. A puzzle for an octopus might involve opening a series of latches to reach crabmeat. An object such as a pill bottle may become a toy. These activities show how smart octopuses are. "Octopuses play," says scientist Jennifer

Mather, "and play is something that intelligent creatures do."[7]

RESEARCHING SMART ANIMALS

Scientists have learned a lot about smart animals. They have observed animals using their intelligence in the wild. People have seen animals use tools, communicate, and work together. Scientists have also studied animal smarts. Using games and tests, scientists have explored how animals learn and think.

But there's much scientists still don't know about animal intelligence.

People continue to watch, test, and learn.

Smart animals may have a lot more tricks to show humans. And some animals that don't seem very intelligent now may surprise scientists in the future.

GLOSSARY

biology

the study of plants, animals, and other forms of life

camouflage

to hide by blending into an environment

captivity

the condition of being held by humans

culture

a shared way of doing things

facial expressions

looks on faces that communicate something

gestures

body movements that communicate something

mimicking

copying someone or something else

nerve cells

the main parts that make up the body system that allows animals to move and sense their surroundings

species

a closely related group of living things that can produce fertile young

SOURCE NOTES

CHAPTER ONE: CHIMPANZEES

1. Quoted in Victoria Gill, "Chimps Show Off Their 'Signature' Drum Beats," *BBC News*, September 7, 2022. www.bbc.com.

CHAPTER TWO: AFRICAN GRAY PARROTS

2. Quoted in Laura Bond, "Animal Behaviorist Irene Pepperberg to Present Parrot Research," *University of Tennessee at Chattanooga*, March 23, 2012. https://blog.utc.edu.

3. Quoted in Justin Saglio, "When a Bird Brain Tops Harvard Students on a Test," *Harvard Gazette*, July 2, 2020. https://news.harvard.edu.

CHAPTER THREE: BOTTLENOSE DOLPHINS

4. Quoted in Nell Greenfieldboyce, "Dolphins Learn Foraging Tricks from Each Other, Not Just from Mom," *NPR*, June 25, 2020. www.npr.org.

5. Denise Herzing, "Do Dolphins Have a Language?" *NPR*, June 26, 2020. www.npr.org.

CHAPTER FOUR: OCTOPUSES

6. Quoted in Brandon Keim, "Tool Use Found in Octopuses," *Wired*, December 14, 2009. www.wired.com.

7. Quoted in Brendan Borrell, "Are Octopuses Smart?" *Scientific American*, February 27, 2009. www.scientificamerican.com.

FOR FURTHER RESEARCH

BOOKS

Michelle Garcia Andersen, *Animal Problem-Solving*. Vero Beach, FL: Discovery Library, 2022.

Anita Ganeri, *Jane Goodall: A Life with Chimps*. New York: Random House, 2019.

Sam Hume, *An Anthology of Aquatic Life*. New York: DK, 2022.

INTERNET SOURCES

"Can Parrots Really Talk?" *Wonderopolis*, August 15, 2019. https://wonderopolis.org.

"Common Bottlenose Dolphin," *National Geographic*, n.d. www.nationalgeographic.com.

Olivia Judson, "What the Octopus Knows," *Atlantic*, January 2017. www.theatlantic.com.

WEBSITES

DK FindOut!: Invertebrates
www.dkfindout.com

DK FindOut! hosts games, articles, and interactive activities about science, art, history, and more. Its "Invertebrates" page contains information about octopuses and other fascinating creatures.

San Diego Zoo Wildlife Alliance
https://animals.sandiegozoo.org

The San Diego Zoo Wildlife Alliance is a nonprofit organization dedicated to conserving plants and animals. Its website includes educational resources and live videos of the animals at the San Diego Zoo in San Diego, California.

Whale and Dolphin Conservation
https://us.whales.org

Founded in 1987, Whale and Dolphin Conservation was created to protect whales and dolphins and research these smart creatures. Its website contains information about different species of dolphins as well as ways to protect these clever animals.

INDEX

IMAGE CREDITS

Cover: © Rich Carey/Shutterstock Images
5: © madcorona/iStockphoto
7: © Norma Cornes/Shutterstock Images
9: © NaluPhoto/iStockphoto
11: © Gerdie Hutomo/Shutterstock Images
12: © guenterguni/iStockphoto
16: © Ari Wid/Shutterstock Images
18: © BigDane/Shutterstock Images
23: © Andreas Krone/Shutterstock Images
25: © GaevoyB/Shutterstock Images
28: © Passakorn Umpornmaha/Shutterstock Images
31 (gorilla, bonobo): © ALX1618/Shutterstock Images
31 (African gray parrot): © Noch/Shutterstock Images
31 (dog): © Artco/Shutterstock Images
35: © gilkop/Shutterstock Images
37: © Nicolas Sanchez-Biezma/iStockphoto
39: © Redders48/iStockphoto
41: © Anita Kainrath/Shutterstock Images
42: © Tory Kallman/Shutterstock Images
47: © Henner Damke/Shutterstock Images
49: © Vittorio Bruno/Shutterstock Images
51: © SergeUWPhoto/Shutterstock Images
52: © Andrea Izzotti/Shutterstock Images
57: © Vladimir Wrangel/Shutterstock Images

ABOUT THE AUTHOR

Clara MacCarald is a freelance writer with a master's degree in ecology and natural resources. She lives with her family in an off-grid house nestled in the forests of central New York. When not parenting her daughter, she spends her time writing nonfiction books for kids.